COUNTDOWN TO SPACE

COUNTDOWN TO SPACE

SATURN—
The Sixth Planet

Michael D. Cole

Series Advisors:
Marianne J. Dyson
Former NASA Flight Controller
and
Gregory L. Vogt, Ed. D.
NASA Aerospace Educational Specialist

Enslow Publishers, Inc.

40 Industrial Road	PO Box 38
Box 398	Aldershot
Berkeley Heights, NJ 07922	Hants GU12 6BP
USA	UK

http://www.enslow.com

Library of Congress Cataloging-in-Publication Data

Cole, Michael D.
    Saturn : the sixth planet / Michael D. Cole.
        p. cm. — (Countdown to space)
    Includes bibliographical references and index.
    ISBN 0-7660-1950-0
    1. Saturn (Planet)—Juvenile literature. [1. Saturn (Planet)] I. Title. II. Series.
QB671 .C65 2002
523.46—dc21

                                                       2001006706

Printed in the United States of America

10 9 8 7 6 5 4 3 2 1

To Our Readers: We have done our best to make sure all Internet addresses in this book were active and appropriate when we went to press. However, the author and the publisher have no control over and assume no liability for the material available on those Internet sites or on other Web sites they may link to. Any comments or suggestions can be sent by e-mail to comments@enslow.com or to the address on the back cover.

Photo Credits: Craig Attebery (Jet Propulsion Laboratory), p. 38; Reta Beebe (New Mexico State University), D. Gilmore and L. Bergerson (Space Telescope Science Institute), NASA, p. 7; Heck's Pictorial Archive of Art and Architecture, p. 9; History of Science Collections, University of Oklahoma, p. 11; JPL/NASA, pp. 15, 17, 18–19, 22, 24, 26, 27, 31, 34 (top and bottom), 36; Bob Kanefsky, NASA Ames Research Center, p. 4; Lunar and Planetary Institute (LPI), pp. 13, 14, 29; NASA/STScI/AURA, pp. 21, 42; U.S. Geological Survey Flagstaff, NASA, p. 34 (center).

Cover Photo: NASA (foreground); Raghvendra Sahai and John Trauger (JPL), the WFPC2 science team, NASA, and AURA/STScI (background).

CONTENTS

Saturn's rings make the planet unique and beautiful. This image compares the relative sizes of Saturn and Earth. Two of Saturn's moons are also visible at the correct size and distance from the planet: Rhea to the far left and Dione at the bottom.

I

The Saturn Storm

Seeing Saturn through a telescope for the first time is an exciting experience. No matter how many pictures of Saturn a person has seen, it is still a shock to see the planet for real. It somehow seems hard to believe that those rings are *really there*. But they are. It is not hard to imagine Galileo's surprise at seeing the strange "handles" on Saturn when he first looked at the planet through his primitive telescope nearly four hundred years ago.

More recently, in September 1990, a number of amateur astronomers, who were observing Saturn through their telescopes purely for enjoyment, got a big surprise. Closely studying their view of Saturn, they saw what appeared to be a large white spot near the planet's equator.

Those who saw the spot on Saturn contacted professional astronomers at astronomical observatories. After learning of the spot, the professional astronomers turned the larger telescopes in their observatories toward the planet. The images in the larger telescopes confirmed that something special was happening on Saturn: An enormous storm extended completely around the planet.

Nothing like it had been seen on Saturn since 1933, when a similar spot appeared. Telescopes in that era did not allow scientists to study the event in as much detail as was now possible with more modern instruments. Because scientists knew the storm was probably a once-in-a-lifetime event, a scientific team was organized to use the Hubble Space Telescope to observe the storm. The Hubble Space Telescope orbits Earth, high above its atmosphere. It is able to produce sharp, clear images of distant objects in space because it does not have to look through Earth's blurring atmosphere, as do telescopes on the ground. A special team of scientists and engineers reprogrammed the telescope's schedule of observations. Observing the storm on Saturn became the top priority.[1]

In mid-November 1990, the Hubble Space Telescope was able to observe the storm on Saturn for several days. When Hubble's pictures were put together in a series, they produced a movie of the storm's motion around Saturn's equator. Scientists studying the movie saw that winds were moving through the storm far faster than the speed of sound on Earth, at more than 1,000 miles

The Hubble Space Telescope captured a rare storm on Saturn near the equator. The width of this storm is equal to the diameter of Earth.

per hour. The pictures also gave scientists a better understanding of what the clouds were made of and how deep they were within the atmosphere.[2] They speculate that the cloud was ammonia crystals. It may have been as thick as 1,200 miles, which is about the distance from San Diego, California, to Seattle, Washington.

Scientists were learning something new about Saturn, but only because the amateur astronomers had alerted them to what was happening there. The amateur astronomers who turned their telescopes to Saturn and saw the giant storm were only the latest to observe something grand about the planet. With its complex system of rings and many moons, Saturn is a planet unlike any other in the solar system. People have been looking to it with wonder for a very long time.

2

The Ringed Wonder

Saturn is the farthest planet from Earth that can easily be seen in the night sky with the naked eye. It has been known to the human race since the earliest civilizations, possibly even three thousand years ago.

Many cultures such as the Babylonians, ancient Greeks, and ancient Chinese recognized the "wanderers" that moved across the background of stars. The stars stayed in their same patterns and positions year after year, century after century. But these moving objects shifted slowly from one pattern of stars to another as the months and years passed. These objects included the planets Mercury, Venus, Mars, Jupiter, and Saturn, as well as the Sun and the Moon. The word *planet* comes from the Greek word *planetes*, which means "wanderer."

Uranus is not easily visible to the naked eye, and telescopes are required to view Neptune and Pluto: These three planets were not discovered until much later, after the invention of the telescope.

Ancient cultures named the five visible planets after gods in their mythology. The ancient Greeks associated the planets with their gods Hermes, Aphrodite, Ares, Zeus, and Cronos. The Romans later had their own version of the Greek gods and gave the five planets the names we know today. In Roman mythology, Saturn was the god of agriculture. Saturn is also the root of the English word *Saturday*.[1]

Saturn is not really a wandering star. It is a planet in orbit around the Sun. But it orbits at a far greater distance from the Sun than Earth does. Earth orbits the Sun at a distance of 93 million miles

The Romans named the planet after their god of agriculture, Saturn.

(149 million kilometers). Saturn orbits at 887 million miles (1.4 billion kilometers) from the Sun. Depending on where Earth and Saturn are in their orbits, Saturn can reach a distance of more than one billion miles (1.7 billion kilometers) from Earth, and it never comes closer than 741 million miles (1.2 billion kilometers).[2]

Saturn's orbit is 9.5 times more distant from the Sun than Earth's orbit. Because it is so far from the Sun, Saturn's orbit around the Sun takes much longer than Earth's. It takes more than 29 Earth-years for Saturn to complete one orbit around the Sun. That is a long time to wait for each birthday.

Saturn in Early Astronomy

In 1610, Italian astronomer Galileo Galilei became the first person ever to observe Saturn with a telescope. Although the lenses of his telescope lacked today's quality, Galileo immediately noted the planet's odd appearance. In Galileo's telescope, Saturn appeared to have something that looked like handles on either side of the planet. For almost fifty years, early astronomers were unsure what caused Saturn to appear this way. This was partly because Saturn's rings, as viewed from Earth, were tilted at an angle that made the rings difficult to observe clearly.

In 1655, Dutch astronomer Christiaan Huygens used slightly improved telescope lenses to study Saturn. He stated that the planet was surrounded by a solid ring.

Two hundred years later, British physicist James Clerk Maxwell made a different claim. He believed that the forces and tensions caused by the planet's gravity and rotation would have caused such a ring to be broken into many small pieces. Therefore, Maxwell suggested, it was more likely that the rings were made of many tiny

Galileo Galilei lived from 1564–1642. He was the first person to see Saturn through a telescope.

meteorites, freely circling the planet. But Maxwell and other scientists at that time could only guess at what the rings were actually made of.[3]

The Shape of Saturn

Saturn, the second largest planet in the solar system, is about 75,000 miles (120,000 kilometers) wide. That is the diameter of the planet's sphere and does not include the rings. If Saturn were placed right next to Earth, it and its rings would take up the space between our planet and the Moon. And if Saturn were hollow, 844 Earths could fit inside it.

Saturn rotates on its axis very rapidly, making a "day" on Saturn just about eleven hours. The huge planet, rotating at this great speed, forces the regions near Saturn's equator to bulge outward. The diameter at Saturn's equator is about 3,700 miles (6,000 kilometers) wider than the distance between its north and south poles.[4]

Saturn bulges outward this much because it is made almost entirely of gas. It is one of the gas giant planets, along with Jupiter, Uranus, and Neptune. The other planets in our solar system—Mercury, Venus, Earth, and Mars—are made of rock, while Pluto is thought to be a combination of rock and ice. Similar to the other gas giant planets, Saturn is made mostly of hydrogen in different forms. It is about 96 percent hydrogen and 3 percent helium, with traces of other gases such as

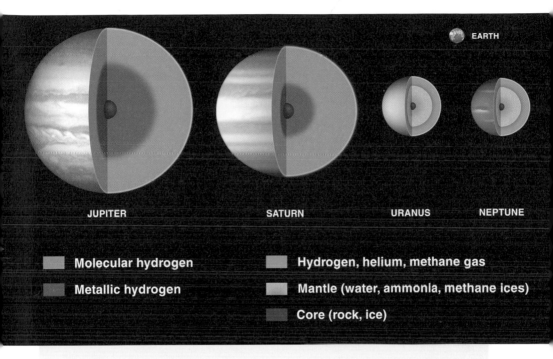

EARTH

JUPITER   SATURN   URANUS   NEPTUNE

■ Molecular hydrogen   ■ Hydrogen, helium, methane gas
■ Metallic hydrogen   ■ Mantle (water, ammonia, methane ices)
■ Core (rock, ice)

The four gas giant planets—Jupiter, Saturn, Uranus, and Neptune—are shown in size comparison to Earth. Saturn is made mainly of hydrogen.

methane, ammonia, and water. These elements all exist as tiny bits of frozen gas in Saturn's atmosphere.

Scientists believe that deep inside the planet, probably about 19,000 miles (30,000 kilometers) beneath the cloud tops visible from space, increasing pressure changes the hydrogen to a form in which it is similar to a metal. This metallic hydrogen forms a giant inner layer. At the planet's center, a solid core made of heavy minerals, ices, and other elements may exist.[5]

Despite Saturn's enormous size, its vast collection of

gases is not very dense. Saturn is the only planet in our solar system with a density that is less than water. This means that if you could find an ocean big enough, Saturn would float in it.[6]

Saturn's Rings

The most distinctive feature of Saturn is its system of rings. The rings are made mostly of water ice and rocky particles coated with ice. These icy particles reflect light very well. That is the reason the rings are so easy to see with telescopes from Earth. Jupiter, Uranus, and

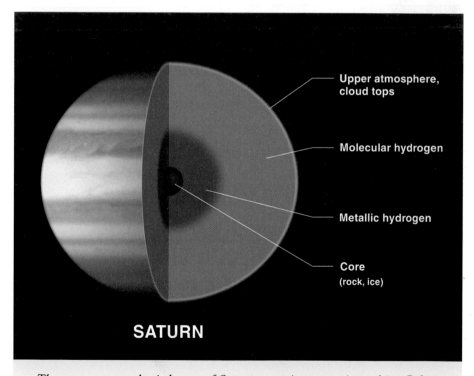

SATURN

The upper atmospheric layers of Saturn contain ammonia and ice. Below this are layers of hydrogen. The core is thought to be made of rock and ice.

Neptune also have rings, but they are not vast systems of rings like those that encircle Saturn. They are made of less reflective particles, making them invisible from Earth.

Saturn's rings stretch across a distance of more than 298,000 miles (480,000 kilometers)—a size greater than the distance between Earth and the Moon. Although the distance covered by the rings around Saturn is huge, the rings themselves are very narrow. The width of the ring

Saturn's stunning rings, which are made mostly of ice and rock, make Saturn a unique place in our solar system.

particles orbit an area around the planet of only about one tenth of a mile. And the particles together do not amount to much. If all the particles in the rings could be clumped together into one body, such as a moon, that moon would be no more than 70 miles (112 kilometers) wide.[7]

How Did the Rings Get There?

Three theories attempt to account for the existence of Saturn's rings. One theory assumes that the ring particles are material left over from the formation of the solar system. The particles could not form into a moon because they were trapped too close within the planet's gravity. At this distance from Saturn, the planet's gravity and its rapid rotation created forces that would not allow a moon to clump together and take shape.

The second theory explains that an earlier moon of Saturn ventured too close to the planet. This moon was then pulled apart by the same effects of the planet's gravity and rotation. The many pieces of this fractured moon were then slowly spread around the planet, forming the rings we see today.

The third theory is that a moon of Saturn was struck violently by another object in space. The object could have been another of Saturn's moons, a comet, or a large asteroid. The collision of the two bodies shattered the moon and the impacting object. Their debris eventually formed the rings.

Scientists have different theories about the formation of Saturn's rings. The color variation in this photo is due to the different chemicals that make up the rings.

Most astronomers believe that these three theories for the rings' origin are equally possible.[8] The *Cassini* spacecraft, which will arrive at Saturn in 2004, may discover new evidence about the planet's rings that will lead scientists to favor one of the theories or perhaps force them to adopt a new one.

Saturn is like a mini solar system, surrounded by a number of orbiting moons. These moons have a great variety and are just as amazing as the planet.

SATURN[9]

Age
About 4 billion years

Diameter at equator
75,000 miles (120,000 kilometers)

Diameter at the poles
68,000 miles (108,000 kilometers)

Planetary mass
95 Earth masses

Distance from Sun
887 million miles (1.4 billion kilometers)

Closest passage to Earth
741 million miles (1.2 billion kilometers)

Farthest passage from Earth
1 billion miles (1.7 billion kilometers)

Orbital period (year)
29 years, 167 days

Rotation period (1 day)
10 hours, 39 minutes

Average Temperature
-202°F (-130°C), increases with depth

Composition
Solid or molten rock mineral core (scientists unsure), layer of metallic liquid hydrogen, vast outer layer of liquid hydrogen, mostly hydrogen gas atmosphere

Atmospheric composition
96 percent hydrogen, 3 percent helium, traces of other gases

Wind speeds
1,100 miles (1,770 kilometers) per hour at equator

Gravity
92 percent of Earth's gravity

Number of known moons
30 (and there are probably more)

Composition of rings
Small bits of ice and ice-covered dust and rock

Diameter of rings
About 298,000 miles (480,000 kilometers) from outer edge to outer edge

Amount of heat received from Sun
About 1 percent of the amount Earth receives

3

Saturn's Moons

Saturn is orbited by thirty known moons. Four of those moons were discovered in the year 2000, and more may be found when the *Cassini* spacecraft reaches Saturn. The first moon of Saturn to be discovered, by Christiaan Huygens in 1655, was Titan.

Titan was discovered first because of its size and brightness. This golden orange moon is more than 3,200 miles (5,150 kilometers) wide. It is larger than the planet Mercury and is the second largest moon in the solar system. Only Jupiter's moon Ganymede is larger. Titan appears bright in a telescope because it is surrounded by a thick atmosphere that reflects a great deal of light. Its atmosphere is made mostly of nitrogen, with traces of methane and other gases. Titan is the only known moon in the solar system that has a thick atmosphere.[1]

Earth's atmosphere is made mostly of nitrogen as well. But unlike Earth, Titan's atmosphere contains no oxygen. The atmospheric pressure at the surface is 1.6 times greater than Earth's. The temperature on the surface of Titan plunges to -289°F (-178°C).[2] Such frigid conditions make an environment very hostile to human life. But from their studies of Titan, scientists can piece together a picture of an unusual moon with clouds, rains, and seas.

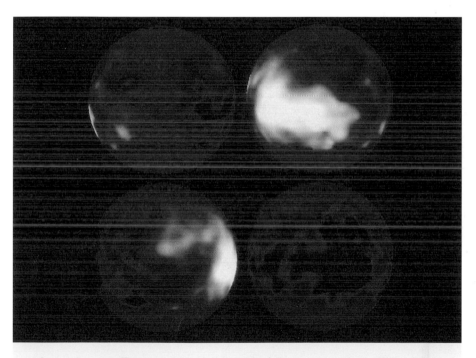

Saturn's moon Titan is shrouded in a thick hazy atmosphere. This photo, taken by the Hubble Telescope, shows a bright surface feature the size of Australia.

"Standing on the surface of Titan, we would see a very dimly lit world, as bright as Earth under a full moon," said professor Caitlin Griffith of Northern Arizona University. "Below the orange sky, the Sun would appear as a diffuse light source through Titan's high smog. At night we would not see stars through the smog's veil. . . . Every week, sparse clouds would appear below the orange haze but still high in the sky, barely visible. They would quickly produce rain and disappear."[3]

The moons Mimas, Enceladus, Tethys, Dione, and Rhea appear to be made mostly of water ice. They range

Photos of Saturn and its moons Dione, Tethys, Mimas, Enceladus, Rhea, and Titan were grouped for this composite picture.

in size from Rhea, at 949 miles (1,528 kilometers) wide, to Mimas, at 244 miles (392 kilometers). These moons appear to be billions of years old. Mimas, Tethys, Dione, and Rhea are densely covered with craters. Enceladus has a newer surface and therefore few craters. Internal heating from its core may have caused interior ice to melt, cracking the icy surface and allowing new liquid to flow onto the surface and form new ice.

The most spectacular crater on these moons is on the surface of Mimas. The crater Herschel, at 80 miles (130 kilometers) wide, covers nearly one third the diameter of Mimas. It is about 6 miles (10 kilometers) deep, with a central peak that rises at least 3.7 miles (6 kilometers) high. Mimas must have barely survived the impact of the object that caused the crater. Fractures caused by the impact's shockwave traveling across the moon are visible all over its surface. If the object that caused the crater had been just a little larger, Mimas would have been broken into several pieces.[4]

Images of Mimas appeared in many newspapers worldwide when the crater was discovered in 1980. But it did not make the newspapers for reasons of science. It appeared instead because the large crater and its central peak gave Mimas an appearance similar to the Death Star made famous by the movie *Star Wars*.

Some of Saturn's moons orbit at less than 50,000 miles (80,000 kilometers) from the planet. Earth's Moon, by comparison, orbits our planet at an average distance

Saturn's moon Mimas is covered with craters.

of about 240,000 miles (384,000 kilometers). The small moon Pan orbits inside a gap between two of Saturn's rings, only 45,000 miles (73,000 kilometers) from the planet's outer atmosphere. Pan takes only 13.8 hours to orbit Saturn, while the moon Phoebe, at a distance of more than 8 million miles (13 million kilometers) from Saturn, takes one and a half years to complete one orbit. The most distant moon of Saturn discovered so far orbits at a distance of 9 million miles (15 million kilometers) from the planet.[5] That is more than 37 times farther away than our own Moon orbits Earth.

Saturn's great distance from Earth limits how much can be learned about the planet through telescopes. Much of our detailed knowledge of Saturn was gained by the voyages of three important spacecraft.

4

Exploring Saturn

The spacecraft *Pioneer 11* was launched by the National Aeronautics and Space Administration (NASA) in 1973. For six years, *Pioneer 11* traveled through space. It passed through the asteroid belt that exists between the orbits of Mars and Jupiter and was the second spacecraft, after *Pioneer 10*, to fly by Jupiter. In September 1979 *Pioneer 11* became the first spacecraft to arrive at Saturn. The spacecraft passed within 12,000 miles (19,000 kilometers) of the planet's cloud tops, giving scientists their first close-up look at the ringed planet. *Pioneer 11*'s cameras allowed scientists to see the true complexity of Saturn's rings for the first time.

Pioneer Studies the Rings

The planet's rings are identified by a series of letters. The A and B rings are easy to see by telescope from Earth.

They are separated by a dark gap called the Cassini division, named after French astronomer Jean-Dominique Cassini, who discovered the gap in 1676. The dimmer C and D rings orbit in an area closer to the planet, inside the orbits of the B and A rings.

The D ring is closest to the planet, starting about 4,400 miles (7,000 kilometers) above Saturn's cloud tops. The main ring system then continues outward across rings C, B, and A, with the outer edge of the A ring 50,000 miles (80,000 kilometers) from the planet. Far

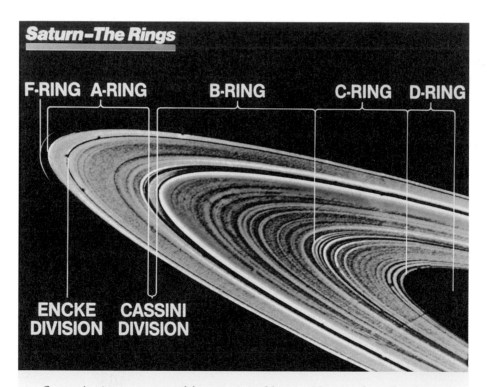

Saturn—The Rings

F-RING A-RING   B-RING   C-RING D-RING

ENCKE   CASSINI
DIVISION DIVISION

Saturn's rings are named by a series of letters. Viewers from Earth can easily see rings A and B with a telescope. Rings C and D are dimmer and orbit closer to the planet.

outside the traditional ring system is the very thin E ring. Its sparse, scattered particles orbit Saturn at a distance of more than 105,000 miles (170,000 kilometers) from the planet.[1]

Pioneer 11 discovered a slender F ring just 2,500 miles (4,000 kilometers) from the outer edge of the A ring. The spacecraft showed that the F ring had a knotted or braided appearance in some places. Its clumpy composition hinted that something was acting upon the ring to cause it to bunch up. Scientists predicted that an undiscovered moon, or more than one moon, was orbiting through this region, causing disruptions in the ring.[2]

The Voyagers

Pioneer 11 was a flyby mission, its path carrying it onward to the outer reaches of the solar system. Just months after the spacecraft left Saturn, astronomers discovered the small moons Atlas, Prometheus, and Pandora orbiting near the F ring. Their existence was confirmed when the

Voyager 1 *helped confirm that Saturn's outer F ring appears braided because of three moons. The moons orbit near the F rings. Their gravity pulls on the particles and causes them to bunch.*

spacecraft *Voyager 1* arrived at Saturn in November 1980. The moons were referred to as "shepherding satellites" because of the way their gravitational effects herded the particles of the F ring into bunches. Although *Voyager 1* was a flyby mission like *Pioneer 11*, it was equipped with a far better set of instruments than *Pioneer 11* and yielded much more scientific information about Saturn.

Following *Voyager 1*'s flyby, *Voyager 2* arrived at Saturn in August 1981. The two Voyager spacecraft sent back thousands of images. They clocked winds near the planet's equator at about 1,100 miles (1,770 kilometers) per hour, far faster than the speed of sound on Earth. The spacecraft's images showed scientists what are called jet stream winds. The existence of these kinds of winds in the planet's upper cloud layer indicates that winds also exist deeper within the planet. The winds are probably present at least 1,200 miles (2,000 kilometers) below the cloud layer.[3]

Voyager 2 penetrated Saturn's upper atmosphere with a radio beam and measured temperatures at different depths within the planet. High in the atmosphere, a frigid temperature of -312°F (-191°C) was recorded. Deeper within the atmosphere, the temperature was a bit warmer, at -202°F (-130°C)—still far colder than the coldest temperatures recorded on Earth in Antarctica.

These temperatures seem very cold, yet Saturn produces more heat than it receives from the Sun. One of the great questions about Saturn is how it could produce this heat.

In Search of Saturn's Heat

Gravity causes Saturn's huge mass of hydrogen and other material to slowly sink closer toward the planet's center. Scientists believe this sinking process squeezes the mass of hydrogen into a smaller volume, resulting in a release of heat. This release of heat causes the temperature within the planet to rise. Although the temperatures in Saturn's upper atmosphere are very cold, temperatures deep within the planet are over 13,700°F (24,740°C).

This process alone cannot fully account for Saturn's producing 1.7 times as much heat as it receives from the

Saturn is a chilly planet compared to Earth. The average temperature is −202°F (−130°C).

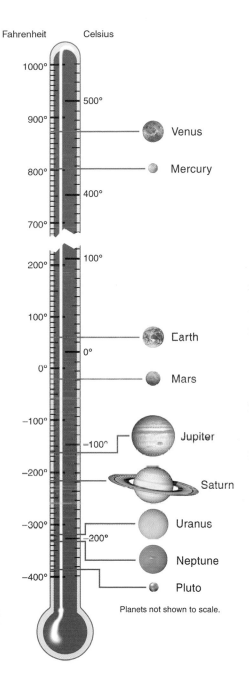

Planets not shown to scale.

Sun. Scientists expected Saturn's atmosphere to be very similar to Jupiter's, which is also mostly hydrogen, with 11 percent helium. But *Voyager 1* discovered that Saturn's atmosphere had only 3 percent helium. Scientists believe this smaller amount of helium is evidence that Saturn's helium is sinking through the planet's hydrogen, releasing additional energy as it sinks. This process may account for the extra heat produced within the planet.[4]

Spinning Storms on a Rapidly Spinning Planet

Most images of Saturn before the Voyagers showed Saturn's cloud bands as very straight and dull in color compared to the more colorful cloud bands on Jupiter. But images from the two Voyager spacecraft revealed more color differences in the cloud bands and showed that much mixing occurs between the bands.

Images also showed large cyclonelike storms whirling through the cloud bands. These oval storms lasted for days or weeks, traveling around the planet many times.

Both spacecraft measured Saturn's rotation to be exactly 10 hours, 39 minutes, and 24 seconds.[5] A day on the much smaller Earth is more than twice as long.

The Moons Up Close

The Voyagers focused some attention on Titan, taking pictures of and measuring the moon's surface temperature and atmospheric pressure. Scientists thought Titan was the largest moon in the solar system. But measurements by the Voyager spacecraft showed

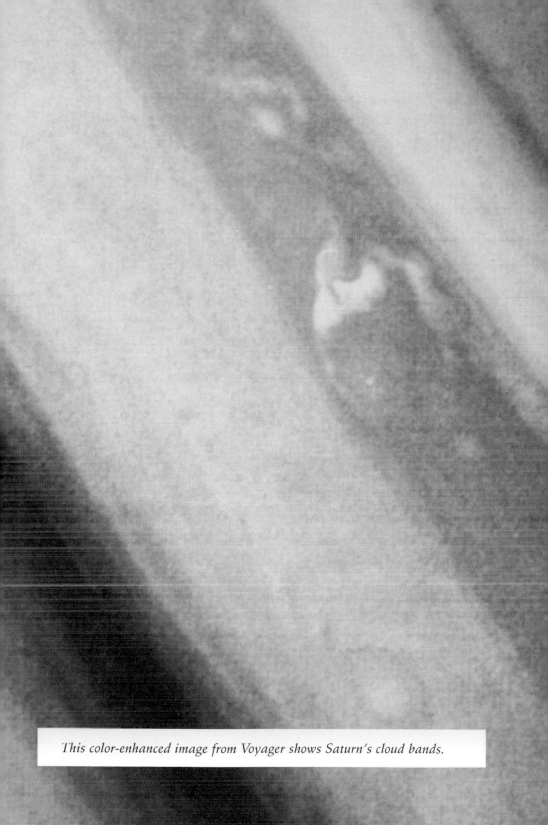

This color-enhanced image from Voyager shows Saturn's cloud bands.

Titan's atmosphere to be more than 250 miles (400 kilometers) deep. This meant that the moon itself was slightly smaller than Jupiter's largest moon, Ganymede.

The near-infrared camera aboard the Hubble Space Telescope was able to penetrate the atmosphere of Titan. Images from the camera show the blurry outline of a huge "continent" on the hemisphere of Titan that faces forward in its orbit. The Hubble images do not prove that liquid seas exist, only that there are large bright and dark regions on its surface. The images from Hubble were used to plan the site where the *Cassini* spacecraft's *Huygens* probe will land on Titan.

Huygens will also record data about the makeup of Titan's atmosphere. Many scientists believe the chemistry in Titan's atmosphere may be similar to what Earth's atmosphere was several billion years ago.

The moon Enceladus was found to have many different types of terrain. This indicates that volcanoes have been active on the moon in the last few hundred million years. Some terrain was cratered, while other areas were smooth from recent lava flows. Its many faults, valleys, and new lava flows show that Enceladus has the most active surface of any known moon in the solar system.[6]

Voyager 1 took images of Mimas and its huge crater, Herschel, that gives the moon its Death Star appearance. Mimas may also have geysers erupting from its surface.

Another unusual moon is Iapetus. Images from

Voyager 1 revealed that one side of Iapetus has a bright yellow-white surface and is densely covered with craters. The other side is much darker. The spacecraft found that the bright side faces forward in the direction of Iapetus's orbit around Saturn, while the darker side is the rear or trailing side of the moon. Like Iapetus, most moons of Saturn keep the same side facing toward the planet at all times.

The moon Dione also shows some strange features. On one hemisphere there is a network of bright streaks over a darker surface. Few craters are visible, and the streaks overlay the craters, indicating that the streaks are newer than the craters. The other hemisphere is more heavily cratered and uniformly bright. Scientists believe that after new ice erupted all over Dione's surface to cause the streaks, the moon experienced a period of smaller and less intense meteor impacts. Much of these later impacts occurred on the leading hemisphere of Dione, covering the streaks on that hemisphere but leaving those on the trailing hemisphere undisturbed.

Voyager 2 recorded images of the unusual moon Phoebe. Phoebe orbits the planet in the opposite direction of all the other moons of Saturn. It is also very red. Scientists believe that Phoebe, which is only 135 miles (220 kilometers) wide, is an asteroid that was long ago captured by Saturn's gravity. Phoebe and another moon, Hyperion, do not always keep the same face toward Saturn.[7]

Saturn's moons include Enceladus, Iapetus, and Dione (top, center, bottom). Enceladus has volcanoes. Iapetus has a bright, heavily cratered icy terrain and a dark terrain. Dione has one heavily cratered half and one dark half with bright streaks.

Cassini

Scientists learned much from the Pioneer and Voyager spacecraft, but an in-depth study of Saturn can only be accomplished by a long-term orbital mission to the planet. On October 15, 1997, the spacecraft *Cassini* was launched from Cape Canaveral. It was named after the man who discovered the dark division in Saturn's rings in 1676.

The spacecraft has not been traveling directly to Saturn. Its flight path has carried it through the inner solar system for two swings around the planet Venus, back around Earth, and outward to Jupiter; it is finally moving onward to Saturn. Each of these swings around the other planets allowed the spacecraft to use that planet's gravity for a slingshot effect that increased the spacecraft's speed, sending it outward from the Sun and toward Saturn. These slingshot effects allowed the spacecraft to be launched far more cheaply than a direct route would have cost. They eliminated the need for expensive amounts of extra fuel that would have been needed to boost the spacecraft straight to Saturn. *Cassini* is scheduled to arrive at Saturn in July 2004.

Cassini is about the size of a thirty-passenger school bus and weighs six tons. It will take at least 30,000 color pictures of Saturn, its rings, and at least five of its moons during a mission that scientists hope will last four years. Before its main mission is completed, *Cassini* should

On October 15, 1997, Cassini *was launched, beginning its journey to* Saturn.

make at least sixty orbits of Saturn. Scientists hope these orbits will include thirty flybys of Titan.

Cassini will also release the *Huygens* probe, which will descend through the atmosphere of Titan and, it is hoped, land on its surface. The probe will fall through Titan's atmosphere on a parachute, taking measurements of the atmosphere's composition, wind speeds, and pressure. It will relay these measurements to *Cassini*. *Huygens* will take two and a half hours to descend through the atmosphere. If it lands successfully, the probe may continue to send information for thirty minutes from the surface of Titan.[8]

One of the main goals of the *Cassini* mission is to determine the structure and behavior of Saturn's rings in greater detail. Scientists also hope to get close-up views of the surface features of the major moons. With such pictures, they can begin to understand the likely geologic history of each of the moons.

The movement of hydrogen and helium through the planet's interior creates an enormous magnetic field, or magnetosphere, around the planet. This magnetosphere has many complex interactions with the orbits of Saturn's rings and moons. An instrument aboard *Cassini* will touch and measure the flow of particles in this magnetic field, which will tell scientists more about how it affects the space around it.[9]

The spacecraft will study the atmospheric behavior occurring within Saturn's cloud bands. It will look for

5

Observing Saturn

When it is present in the night sky, Saturn is easily visible to the naked eye. Current magazines such as *Astronomy* and *Sky & Telescope* include monthly charts that indicate where Saturn can be found in the sky. The Internet also gives access to many astronomy Web pages that show the positions of Saturn and the other planets, as well as a great deal of other information about astronomy.

Once a person locates Saturn in the sky, an average pair of binoculars will reveal a hint of Saturn's rings. Very clear sky conditions will improve the view, but the low magnification in most binoculars will keep the rings from appearing as distinct from the planet.

A good amateur telescope, with a lens or mirror four inches to eight inches in diameter, will reveal more of the wonders of Saturn. Most good telescopes available today will give excellent views of the planet, its rings, and its largest moon, Titan.

Just as important as a good telescope are sky conditions. Even when the sky appears clear to the naked eye, the movement of air in the atmosphere, brought on by falling temperatures or evening winds, will cause the image in a telescope to blur. An observer must keep his or her eye to the eyepiece, waiting for those moments when the air is still, and the image will become very clear.

At those moments, an observer at the telescope will clearly see the Cassini division in Saturn's rings. On very still, clear nights, a good amount of magnification can be used to make the image of Saturn in the telescope appear larger, showing even more detail on the planet. An experienced observer will be able to see the planet's dim cloud bands. These appear in slightly different shades of gray that cross the planet horizontally. The different shades are very subtle.[1]

The tilt of Saturn's rings as seen from Earth changes from year to year. The rings are sometimes edge-on with our viewpoint from Earth. Because the rings are only about one tenth of a mile thick, they are almost invisible at these times. At other times, when we see the rings at a greater tilt, the view in the telescope will reveal the planet's shadow on the rings behind it. Sometimes a shadow also appears on the planet itself, cast by the rings coming around its front.[2]

As seen from Earth, the tilt of Saturn's rings changes from year to year as the planet orbits the Sun. These Hubble Telescope images were taken from 1996 to 2000.

Although not everyone owns binoculars or a telescope, many communities have a local astronomy club. Astronomy clubs often hold stargazing programs during which the public is invited to look at the night sky through club members' telescopes. Such programs are a great opportunity to observe Saturn and many other wonders of the universe.

Maybe someday, far in the future, all our questions about the planet will be answered. But greater knowledge of this fascinating world will not diminish our awe of Saturn, the ringed wonder of the solar system.

CHAPTER NOTES

Chapter 1. The Saturn Storm
1. NASA Photo Release No. STScI-PR91-04, *Hubble Views Major Storm on Saturn*, January 17, 1991, <http://www.jpl.nasa.gov/saturn/anim1.html>.

2. Ibid.

Chapter 2. The Ringed Wonder
1. Bill Arnett, "Saturn," *Nine Planets*, October 19, 2000, <http://www.seds.org/nineplanets/nineplanets/saturn.html> (January 22, 2001).

2. Michael E. Bakich, *The Cambridge Planetary Handbook* (Cambridge, England: Cambridge University Press, 2000), pp. 234–235.

3. Ibid., p. 237.

4. J. Kelly Beatty, Carolyn Collins Petersen, and Andrew Chaikin, eds., *The New Solar System* (Cambridge, Mass.: Sky Publishing Corporation, 1999), p. 195.

5. Jean Audouze and Guy Israel, eds., *The Cambridge Atlas of Astronomy* (Cambridge, England: Cambridge University Press, 1994), pp. 190–191.

6. Bakich, p. 245.

7. Arnett.

8. Bakich, pp. 240–241.

9. Ibid., pp. 232–253; Beatty, Petersen, and Chaikin, pp. 193–220.

Chapter 3. Saturn's Moons
1. "Fact Sheet: Voyager Saturn Science Summary," *Voyager Projects: Saturn*, May 24, 1995, <http://vraptor.jpl.nasa.gov/voyager/vgrsat_fs.html> (January 22, 2001).

2. Ibid.

3. "NAU Astronomer Detects Clouds in Titan's Atmosphere," *Northern Arizona University Web Page*, October 25, 2000, <http://www.nau.edu/paffairs/titan_discovery.html> (September 4, 2001).

4. Jean Audouze and Guy Israel, eds., *The Cambridge Atlas of Astronomy* (Cambridge, England: Cambridge University Press, 1996), p. 203.

5. Cornell University Press Release, *Successful Satellite-Hunting Team Finds Four New Moons*, October 26, 2000, <http://www.news.cornell.edu/releases/Oct00/Saturn.moons.deb.html>.

Chapter 4. Exploring Saturn

1. Jean Audouze and Guy Israel, eds., *The Cambridge Atlas of Astronomy* (Cambridge, England: Cambridge University Press, 1996), p. 196.

2. Ibid., p. 198.

3. "Fact Sheet: Voyager Saturn Science Summary," *Voyager Projects: Saturn*, May 24, 1995, <http://www.news.cornell.edu/releases/Oct00/Saturn.moons.deb.html> (January 22, 2001).

4. Ibid.

5. Ibid.

6. Audouze and Israel, pp. 204–205.

7. *Voyager Projects*.

8. Sue Keintz, "Saturn Kids Corner," n.d., <http://www.jpl.nasa.gov/cassini/Kids/kidscorner.html> (January 22, 2001).

9. Dr. Edwin V. Bell, "Cassini Information," *National Space Science and Data Center Web Page*, August 18, 1999, <http://nssdc.gsfc.nasa.gov/planetary/cassini.html> (January 22, 2001).

10. Ibid.

11. J. Kelly Beatty, Carolyn Collins Petersen, and Andrew Chaikin, eds., *The New Solar System* (Cambridge, Mass.: Sky Publishing Corporation, 1999), pp. 239–240.

Chapter 5. Observing Saturn

1. Terence Dickinson, *Nightwatch: A Practical Guide to Viewing the Universe*, Third Edition (Buffalo, N.Y.: Firefly Books, 1998), p. 132.

2. Ibid.

GLOSSARY

asteroid—Small body of rock that orbits the Sun. Most asteroids exist in an area called the asteroid belt, between the orbits of Mars and Jupiter.

atmosphere—The outermost layer of gases surrounding an object in space.

atmospheric pressure—The force exerted by the collection of gases that surround a body in space.

cloud band—Cloud formation, such as the ones around Saturn, in which the clouds are separated into individual horizontal areas that each completely encircle the planet.

comet—A collection of frozen gases and dust that orbits the Sun. A comet's orbit may be very elliptical, going far out into the solar system at its ends and very close to the Sun at other times. Nearer to the Sun, solar heat causes the comet to form a tail of gas and dust behind it.

crater—A violently disturbed area on one body created by the impact of another object from space.

equator—An imaginary line that divides the northern and southern hemispheres of a body in space.

magnetosphere—The area above the surface of a star, planet, or moon in which forces caused by the electrical current within that body can be detected.

orbit—The path of one body in space around another body.

ring—Collection of dust, rocks, or ice that orbits a planet.

volcano—A vent in a planet's crust from which molten material from the interior is ejected.

FURTHER READING

Books

Asimov, Isaac. *The Ringed Planet: Saturn.* Milwaukee, Wis.: Gareth Stevens, Inc., 1995.

Booth, Nicholas. *Exploring the Solar System.* Cambridge, England: Cambridge University Press, 1996.

Cole, Michael D. *Hubble Space Telescope: Exploring the Universe.* Springfield, N.J.: Enslow Publishers, Inc., 1999.

Vogt, Gregory. *Jupiter, Saturn, Uranus, and Neptune.* Chatham, N.J.: Raintree-Steck Vaughn, 2000.

Internet Addresses

Arnett, Bill. "Saturn." *The Nine Planets.* October 19, 2000. <http://seds.org/nineplanets/nineplanets/saturn.html>.

Cassini-Huygens Mission to Saturn and Titan. "Welcome to Kids Corner." n.d. <http://saturn.jpl.nasa.gov/cassini/index.shtml>.

Hamilton, Calvin J. "Saturn." *Views of the Solar System.* n.d. <http://www.solarviews.com/eng/saturn.htm>.

Keintz, Sue. "Saturn Kids Corner." *Cassini to Saturn.* n.d. <http://www.jpl.nasa.gov/cassini/Kids/kidscorner.html>.

INDEX